Abel Hernández-Muñoz

AVIARIES, FLIGHT OF FANCY

AF293932

Abel Hernández-Muñoz

AVIARIES, FLIGHT OF FANCY

Initiation in the interesting hobby of breeding ornamental birds.

ScienciaScripts

Imprint
Any brand names and product names mentioned in this book are subject to trademark, brand or patent protection and are trademarks or registered trademarks of their respective holders. The use of brand names, product names, common names, trade names, product descriptions etc. even without a particular marking in this work is in no way to be construed to mean that such names may be regarded as unrestricted in respect of trademark and brand protection legislation and could thus be used by anyone.

Cover image: www.ingimage.com

This book is a translation from the original published under ISBN 978-613-9-43721-4.

Publisher:
Sciencia Scripts
is a trademark of
Dodo Books Indian Ocean Ltd. and OmniScriptum S.R.L publishing group

120 High Road, East Finchley, London, N2 9ED, United Kingdom
Str. Armeneasca 28/1, office 1, Chisinau MD-2012, Republic of Moldova, Europe
Printed at: see last page
ISBN: 978-620-3-22147-3

INDEX

INTRODUCTION

An aviary constitutes a panorama to appreciate beautiful feathered creatures known as ornamental, ornamental or fantasy birds. Their maintenance in optimal conditions constitutes one of the most captivating entertainments to which a person can dedicate his or her free time.

The aviary is recommended for most ornamental bird species commonly kept in captivity. Aviaries allow birds to exercise in flight, which is paramount to the health of many species.

Aviaries should be as large as possible. Make sure that there are no weak points through which birds can escape or, conversely, other individuals can enter.

In many cases, the aviary can be placed outside without any danger to the birds, but in winter they may need extra protection and warmth.

The aviary should have three parts:

- A vast space, open to the outside, where they can fly.
- A shelter where they can find shelter and where they can feed and rest (fix for this purpose several branches that serve as perches).
- A security porch with two doors so that they can easily enter and prevent any escape.

Place the installation on level ground and in a place where the birds can be as quiet as possible. Make sure there are no drafts. Do not install the aviary under trees, as branches could damage the structure of the netting.

THE AVIARY

The location of the aviary is of vital importance. There should be no air currents (drafts). It must be protected from sudden changes in temperature. They should receive the sun only a few hours a day. Special care should be taken in the winter because these birds are tropical and cold damages them. It should be protected from rain, and have good ventilation.

When considering the construction of an aviary for the breeding of several pairs, the following should be taken into account:

The entire premises should be protected with nets so that if a bird escapes from the cage, which happens frequently, it remains inside the aviary and can be recovered, also avoiding the entry of rodents and other vectors that may transmit diseases.

A loose bird is sometimes difficult to catch inside the aviary, for that purpose a jamo or net with the edges of the ring padded or sprinkler with water helps, because if the bird is sprayed with water, then the weight of the feather makes it difficult for the bird to fly.

The aviary must have its faucet (or basin) with water very close, this requirement should not be ignored, otherwise you will have so many inconveniences that you end up abandoning the breeding or making the investment of carrying water near the aviary.

- **The importance of living space.**

It is necessary, at the moment of placing several birds in the same cage, to determine if they are compatible, being preferable only one species per cage, and also to determine if there are too many birds in the same place, which can make them aggressive when competing for food and space. There are very aggressive birds that should be placed in individual cages, adult and juvenile birds should not be placed in the same cage and the possible rivalry of adult males during the breeding season should be assessed.

Knowing that each species of ornamental bird has a different range of living space and tolerance to the invasion of that space by other birds is essential when working in the breeding and marketing of these, if these characteristics are taken into account in the species that we market we will achieve lower economic losses and most importantly the health of the bird will be preserved. It is also very common that within a group of birds there is the tendency of some to pluck the feathers of their companions either to try to use them as nesting material, looking for nutrients in the birth of the feather or simply to draw attention to some coloration of the base of the canyon of the feather. It is very common in light colored canaries in molt that when they see the reddish tone of the blood irrigation to the new feather, they pluck it from the partner causing hemorrhages that stain all the plumage of this one, unleashing then a pecking of other birds on the stained one.

In other species such as psittacine birds, there are cases where out of frustration and stress the bird may try to destroy the plumage of a cage mate and even destroy its own feathers.

THE CAGE

It is not always possible to have an aviary. In that case, choose a spacious cage so that the birds can fly from one perch to another. In the case of canaries or parakeets, a cage of 60x65x30 cm is optimal for a pair.

Make sure that the door closes properly and that safety regulations are observed.

Place the cage on a piece of furniture so that it is at eye level. Do not leave it in the sun or in a place where it is drafty, too close to a heat source or too accessible to a cat or other pet.

- **Cage size.**

The appropriate dimensions of bird cages depend to a large extent on the species to be placed in them and the number of animals of these species, but the dimensions can be very controversial depending on the actual housing needs between one bird and another, so we will refer to the minimum dimensions for the European regulation of cage dimensions and the guidelines for pet stores.

The space between the wires or mesh that make up the cage must prevent the birds from being able to take their heads out of the cage, thus avoiding accidents due to the bird accessing with its beak to places in the vicinity of the cage or getting its head trapped between the wires.

The cages must be made with materials that do not cause intoxication to the birds, preventing them from being coated with paints containing harmful substances; it must be controlled that there are no protrusions, sharp areas or wires with points that can cause damage to the birds; the cages must be easy to clean and have secured doors to avoid escape accidents, especially when species such as psittacines, experts in opening the doors of the cages thanks to the mobility of their beaks, are marketed.

The cages should be placed on pedestals, raised from the floor in visible places and if possible placing parabanes between one cage column and another to avoid birds tending to interact from one cage to another.

In the case of large birds, they should be housed in cages that allow them to open their wings and perform a certain level of exercises, being the dimensions of these cages at least three times the wingspan with a perch in the center allowing the bird to flap its wings at certain times without wearing out its feathers by rubbing against the mesh; While smaller birds, which are generally more lively and active, need larger cages in relation to their body size than larger birds, in other words, the cage/bird cage ratio for this type of bird in the pet store requires more space.

As a result of the disparity of criteria and aware of the importance of living space for birds in captivity, regulations have been developed to establish the dimensions of cages and the number of birds per type in these in pet stores, even regulating the type of cage for sale by species that you want to have as a pet.

When placing perches or perches for small birds in cages, they should be at a prudent distance from the corners or sides of the cage to prevent the bird from turning over and rubbing its tail on the wires and suffering wear and tear to the feathers.

Inside the cages should be placed feeders of sufficient size so that all birds can have access to food and water without having to fight for a space in front of the feeder, in addition to placing toys inside the cages will allow the distraction of the bird in such a confined environment and adapting to this attachment that can then have when purchased as a pet.

In a pet store, hand-raised birds should be separated from those raised by their parents, since the former are closer to humans because they are artificially bred and less likely to defend themselves against aggression from the others, requiring special attention.

HYGIENE

Generally the trays are made of polished and shiny aluminum, make the cleaning in such a way that these characteristics are always maintained (polished and shiny aluminum).

The mesh that serves as the floor of the cages should be free of droppings and the tray underneath should be cleaned **daily**. The floor of the aviary should be swept and at the end of sweeping a bucket of water should be poured in aviaries placed on rooftops or outdoors where the open environment favors the evaporation of water in a short time.

In the case of **passerines**, because of their weak beak, wood can be used. Most breeders have a nest prepared similarly to how the female prepares it and when the one in use becomes very dirty then they change it, although in general this happens only once or at most twice during the whole period.

Leftover food left in the cage can rot and become a source of serious diseases.

Once a year do a general cleaning of the aviary, closing it and disinfecting it with formalin for 24 hours, although this is not a frequent practice, it is recommended in many texts on the subject of aviaries. Actually, everything seems to indicate that this conduct is only recommended under veterinary prescription.

Remember when cleaning the cage trays that it is not enough to clean them with a spatula, also use a small sponge dampened with water. When placing the tray make sure it is completely dry.

Although this does not always need to be done, you should clean the tray and cage floor with chlorinated or formalinized water about once a month or whenever you find signs of diarrhea.

CARE

Always look for a partner, many species need companionship for a long life. Always remember that these precious animals <u>depend on you.</u>

You must be constant in the type of food because changes in the food alter their metabolism and digestion.

Always feed at the same time and buy feed only in the stores provided for this purpose. These birds are essentially granivorous birds with continuous digestion, so the following recommendations are appropriate.

It is recommended that the birds always have food within their reach.

Of perishable food (that does not spoil) more than they consume in a day. Food that is likely to spoil should be removed daily and replaced with fresh food. Once a week provide them with a small piece of fresh fruit.

Place the cage in a cool, bright and cozy place, but protect them from drafts.

Change the water every day, many birds tend to wipe their beaks in the water, and also often droppings appear in the drinker which creates a breeding ground for germs of all kinds.

When cleaning the drinker be sure to remove the slime that forms overnight on the walls and bottom of the drinker.

Keep the cage and tray clean of food debris and droppings that are breeding grounds for mange and other diseases in birds that although **not** transmitted to humans will be very harmful to your birds.

THE CAGES

An aviary should have several different types of cages, according to the characteristics and needs: breeding, flight, reserve, transport, nursing and quarantine cages.

BREEDING CAGES

These cages are used for breeding; in passerines, wooden elements can be used.

FLIGHT CAGES OR CAGES.

They should be much larger especially in height, where we will place the chicks after independence and until they are ready for reproduction and where they will give with free flight, the necessary strength to their wings.

Special care should be taken with overcrowding, which results in a real torture for the birds whose characteristic is to defend their territory, the fight and death will be frequent if we do not respect the vital space they need. In these cages the perches (perches) should be placed at the ends, in such a way that they do not impede the free flight through the cage, being these of different diameters so that the bird can exercise with the toes of the legs, for example use in some the diameter of 1 cm and others 2 cm. If possible have two flight cages, one for each sex.

DIMENSIONS IN BREEDING CAGES AND FLIGHT CAGES

Characteristics in centimeters	Useful dimension in breeding Passeriformes	Useful dimension in the cages
Depth	25	60
Height	30	90
Long	40	150

Characteristics in centimeters	Useful dimension in rearing Psittaciformes	Useful dimension in the cages
Depth	25	60
Height	40	90
Long	60	150

In the cages, the height is of special importance because these birds prefer heights, since this type of bird remains 90% of the time in the highest part of the cage, generally coming down only to eat.

This can be achieved by additionally hanging the cage as high as possible, these cages should be a minimum of 80 to 90 cm. high.

If you are thinking about breeding and reproduction, you should think about assigning a room or an enclosed and roofed space for the aviary.

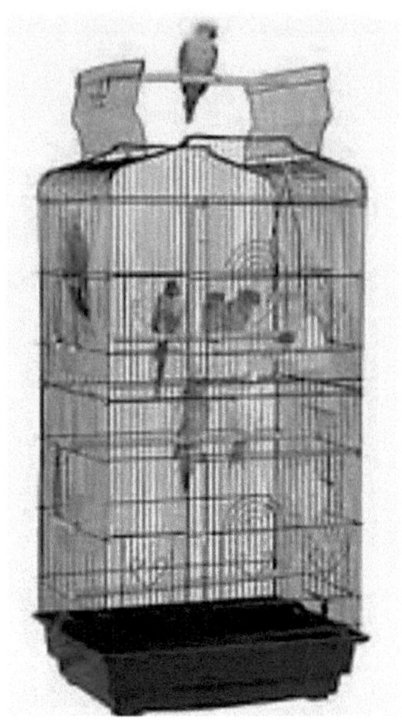

RESERVE CAGES

If you have an aviary with several pairs, you should have some reserve cages to use if the need arises, in general you should have a couple of these cages.

TRANSPORT CAGES.

These cages are used to move the birds to another location.

For example, to take them to the veterinarian. These cages should be small, completely covered on three sides so that the bird does not see out and is not frightened during the trip, the fourth side can be mesh to ensure ventilation.

NURSING CAGES

One of the cages should be used for this purpose, if possible outside the aviary, so as not to contaminate the rest of the aviary.

This shall consist of a heat source The sling bar must be low.

QUARANTINE CAGES.

After a contagious disease or when a new bird comes in from the street, we must keep it in quarantine outside the aviary, as a safety measure, observing it for at least 40 days before incorporating it into the aviary.

Remarks.

The closing of the cages must be secure because the birds, which are very ingenious and persistent, will end up escaping if we trust them, some breeders use as a security measure a clothespin.

The tray supports are traditional but you can modify them (using a pliers or tweezers as a tool) and adapt it in such a way that if you need to put mouse poison in the tray, the poison will not be reached by the beak of your birds.

This is only necessary where rodents are abundant.

NESTING

In **Psittaciformes** the parts of the nest are: the entrance and exit hole, the perch supporting the bird, the inspection window and the reventing holes.

The placement of the nest depends on the distribution of the cages and the cages can be placed in block, parallel or interleaved.

In block. One cage next to the other separated by one or two centimeters and in this case the nest is placed at the front on the opposite side of the cage entrance door.

Parallel. The cages are placed two by two separated from each other by one or two centimeters, but the nests are placed to the right and left of each cage.

As can be easily observed, block placement makes much better use of the available space, but parallel placement offers greater aesthetic beauty when sufficient space is available.

Interleaving. Another way of placement is with the nests intercalated between the two cages, where the pairs do not see their neighbors, avoiding, according to some, unwanted crushes.

For **passerines** the dimensions of the nests are between 12 cm in height, 12 cm in length and 8 cm in width. These are made of wire mesh, wood, plastic or metal and <u>are generally placed inside the cage</u>.

The material to form the nest is jute thread (jute sack) previously boiled, these threads should be cut between 8 to 12 cm. long and placed on the sides of the cage where the female can pick them up with her beak and start building the nest. When the nest is already made, only then a piece of cotton will be placed for its termination.

It is useful to always have a "ready to use" nest made by the breeder and similar to the one made by the mother, so that when we need to sanitize one, we only have to remove it and put in its place the one we have ready.

Breeders who prefer to use the nest outside the cage generally do so with the nests sandwiched between the two cages, where the pairs do not see their neighbors, some say to avoid unwanted crushes.

FOOD

First of all, I would like to start by reminding you that the birds we have **depend entirely on us and it** is a real crime to neglect them. They have no chance of getting their own food and they don't have it because they have lost it through hundreds of generations of foster breeding. We are entirely responsible for this, do not forget it, only by loving them and taking care of them with care is it morally justified to keep them.

Although these birds are capable of eating various foods, we must never forget that they are essentially granivorous.

Birds are animals of continuous digestion, feed supply must take this into consideration, fasting can cause serious digestive disorders and of course poor structural development.

The food must be supplied daily, each day taking care to throw away the chaff which is the remains of the previous meal. This is vital because it has happened that beginners believe that there is food *where there is only straw* and the birds have died of starvation, as many species do not dig the food and the shells will cover the seeds underneath. Usually by blowing on it, the chaff flies away and is removed.

Ferplast Comedero Marrón
para Canarios y Pájaros
Exóticos

TIME REGIME

An advisable dietary schedule would be:

Early morning ....The soft

Midday.. Greens, fruits and vegetables

In the afternoon..................................... All grains

Water is also an indispensable element, and should be changed daily, we assure you that the amount of microorganisms it contains after 24 hours is abundant. Clean well the slime that adheres to the walls of the container.

For the fancier, the mixtures sold in specialized establishments are sufficient and he should always go to them. Buying food for your birds from unscrupulous resellers is a mistake you should not make, since they, in their eagerness to profit, prepare tremendously cheap mixtures that do not meet the necessary requirements.

Trixie Comedero de Acero
Inoxidable con Gancho para
Pájaros

COMPONENTS

Let us now look at the necessary components in the diet: *proteins, lipids, carbohydrates, vitamins and minerals.*

As you have seen, they do not differ in any way from the components of the diet of any living being, therefore we are not going to explain each one of them, which, on the other hand, can be found in any nutrition text.

There are several vegetables (boiled, but not seasoned) that these birds eat without difficulty, although it is not a common practice to use them in our environment. And those that do use them are fed in the midday hours.

As for fruits, they are able to eat many of them, others are not to their liking, but everything seems to indicate that this has a lot to do with what we have accustomed them to.

In any case, the use of vegetables, fruits and green leaves is recommended as a dietary supplement, but always remembering that these birds are essentially granivorous and that grains should always be the fundamental basis of their diet.

Although you will remember that the bird does not live by grains alone.

Purslane, bitter broom and rosemary are green leaves that they eat with great pleasure, it is a widespread criterion that bitter broom has antiparasitic effects for these birds, the truth is that bitter broom when introduced in water before giving them, they rub themselves in it, wetting their plumage with great pleasure, and eat the small flowers with great pleasure.

WATER

Water should be supplied daily, taking care that it is not too chlorinated. Many breeders dissolve vitamins and medicines in the water, but care must be taken with the concentration because of all the water we supply, they only take a small part (change the water daily).

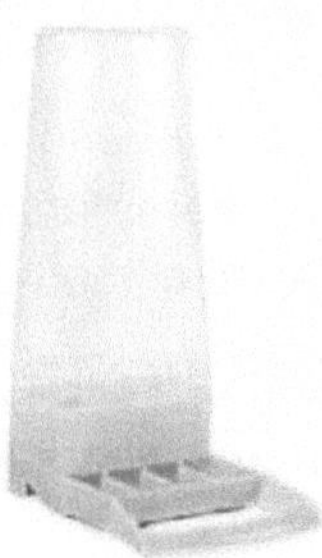

Voltrega Bebedero para Pajarera 430

Arquivet Bebedero Pajaros

BLANDO

The main objective of the soft feed is to provide animal protein and that it does not deteriorate during the day in the conditions of our climate. *It should be fed early in the morning.*

Preparation:

One part boiled and ground egg

Two parts bread crumbs or wheat and corn sift

The soft egg is supplied in a separate well from the beans (some supply the egg alone).

EGG PREPARATION

Boil for 15 minutes and the boiled egg <u>with the shell and all </u>is ground in a meat grinder (the shell is rich in calcium) to this ground egg you can add the following additives:

Vitamins, a small portion of a tablet crushed into a fine powder.

Medicines, if indicated by the veterinarian.

Nutritional supplements, Spirulina for example.

The ground egg quickly disappears from the well when we put the feed in, since it is usually the first thing they eat, which guarantees that the vitamins and medicines we add are not wasted.

IN THE DIET WE MUST AVOID

Supplying milk, actually not few are the fans that make this mistake thinking that milk is an excellent food, but they forget that birds <u>are not mammals </u>and that phylogenetically they do not have the digestive apparatus prepared for its adequate assimilation (they do not possess the enzyme lactose, which mammals have to degrade the sugars present in milk) never supply milk.

Constant changes in the schedule in which food is given to them is detrimental, birds like all animals acquire conditioned reflexes especially associated with food and when we keep the same schedule to provide them with food they are better prepared for their enjoyment and assimilation.

Abrupt changes in diet can lead to digestive disorders and molting, so if you want to introduce a new type of food you should introduce it gradually and progressively.

RECOMMENDED TO INTRODUCE IN THE DIET

First of all, vitamins can be added to the water or to the soft.

The boiled and ground egg should be provided as often as possible, this is a very adequate source of protein for our birds, just

Be careful not to overdo it, one egg can be enough to distribute among 4 or 6 pairs. Eggs should be given daily in larger quantities to breeding birds and those in molt, to the rest it is enough with twice a week, excesses lead to fatty liver, fatness, etc.

FOR FOOD PREPARATION

When grinding the grains, a considerable amount of grain dust is produced, which can damage the respiratory tract, hence the need to sift the material that has been ground. <u>The powder obtained can be used in the soft</u>

In all species this step is mandatory as long as they are supplied with wheat and ground corn.

Wheat and ground corn are mixed in the following proportion: 3 wheat and 1 corn.

FOR PASSERINES

In these species the size and shape of the beak should be taken into consideration, here the size of the grain should be between 2 and 4 mm long and one and a half millimeters wide. Therefore, the grains of corn, wheat, etc. should be ground to the mentioned dimensions and sifted to separate them from the dust.

The so-called "fancy mix" contains all the grains that passerines need for an adequate diet, which **together with the soft feed** (already described) complete the diet of these birds.

Of the fresh fruits, tomato seeds are very popular with them, as well as small pieces of oranges, melon, etc.

In the proportion of seeds, **canaryseed should constitute about 40%.**

CHARACTERISTICS OF SOME CEREALS

Birdseed. It is a grain of pleasant taste for all birds, its protein content is around 13%.

Oats. It has a high fiber content (10%) is pleasing to the palate of the birds but because of its high fiber content should be supplied moderately approximately 5 to 10% of the total to be supplied.

Millo. Its protein value is around 11%.

Corn. It has a high carbohydrate content which makes it an excellent source of energy.

Breadfruit. Protein content is sometimes as high as 13% but it is deficient in tryptophan which is an essential amino acid.

Wheat. Its protein value is between 10 and 15% and it is among the best cereals for poultry.

Soybean. It is rich in oils and proteins (34 to 40 %) and its biological value is very high.

BATH

Hygiene is fundamental to ensure the good health of ornamental birds and bathing is crucial to achieve the goal of raising healthy and beautiful birds. There are several types of baths that will best suit certain species, either because of their size or their hygienic behavior.

REPRODUCTION
PAIRS OR COLONIES.

One of the first questions a fancier will ask is whether to put his birds in cages in pairs or in a colony.

Both possibilities have advantages and disadvantages, we will try to see them separately.

COLONY, ADVANTAGES AND DISADVANTAGES

Colony breeding has the following advantages:

The birds are in almost complete freedom and offer a beauty of their own, besides they can be adorned with natural environments that will increase their beauty.

It reduces the work to be done and the time we have to invest.

Couples are formed more quickly as each bird chooses its mate freely, but this is an advantage but has the disadvantage that we cannot control the offspring that will be born, as there will be no controlled genetic **selection**.

Its disadvantages are:

Diseases can easily spread throughout the colony. There is an increase

in pecking.

The chicks may be attacked by other members of the colony.

PAIR BREEDING

They require more work and time.

Couples take longer to start laying eggs.

But you can control who makes up the couple and therefore the offspring that will be obtained.

PARTNER SELECTION

Although a female and a male of the same species will eventually reproduce and therefore color and characteristics will not determine their mating, the truth is that the random pairing, without taking into account certain criteria, leads in the long run to weaken the genetic heritage of our aviary, so some general concepts should be taken into account at that time.

The "wild" specimen, ontogenetically and phylogenetically, is associated with the genes of greater potential, since they have been fixed through natural selection. In homozygous specimens the offspring is more controllable.

In the mating of mutated specimens, introduce the "wild" in at least the third generation, which will give resistance, physical and genetic strength to the offspring.

To pair up, wait until they are a year old, although some species have been known to mate somewhat earlier (after 8 months).

RESTRUCTURING AND REPLACEMENT OF PARTNERS

Sometimes there is the need or the convenience of undoing a couple because after several months they do not breed, sometimes it happens that an old couple that has had several clutches spend months and do not breed again. It also happens that you want to mate a particular specimen with another specimen and the only specimens you have are mated. You should separate the males and put them in individual cages in a place outside the aviary where they cannot see or hear their old mates for at least 20 days. Only then will you be able to form the desired new pairs, in the same situation you will find yourself in if a member of the pair dies or escapes.

This procedure is highly recommended in many species because if it is not done, it is very possible that the death of a specimen may occur if it is introduced without the proper procedure, in the cage with an unaccustomed companion.

Generally, the reproductive life span with good productivity is 3 years, so it is necessary to replace the couples taking into account this aspect.

Introduce the male first in the breeding cage and keep him alone in it for at least 15 days before forming a pair. After 15 days, introduce the female.

Sometimes it happens that the female we have chosen does not stimulate the male who does not pay any attention to her or even sometimes attacks her, if this happens, change the female.

But what normally happens is that very soon the courtship ritual starts as soon as we have introduced the female in the cage.

In the ritual, the male will ruffle the feathers on his head while approaching her, emitting a particular sound similar to a cooing and will begin to feed her by regurgitation, while the female begins to prepare the nest. This description, although classic, is not exact for all species, which may vary according to the case, but in general the ritual is summarized in the action of the male to please the female so that she will allow the mating.

Generally, mating takes place between the eighth and tenth day after the pair is formed.

SEXUALITY AND SEXUAL DIMORPHISM

In some species males can be differentiated from females depending on certain easily observable characteristics.

For example, the adult male parakeet is identifiable by the blue color of the waxy area on top of the beak, while in the female the color varies from white to carmelite.

In some canaries, as well as in some exotic birds, sexual dimorphism is also observed, as will be seen in the corresponding chapters.

ADIMORPHIC SEXUALITY

In Roseicollis, Personatas, Fisheris and other species we do not find sexual dimorphism; on the contrary, females and males have practically the same morphological characteristics.

Experienced breeders insert the finger into the pelvic orifice because it is larger in the female, but even they often make mistakes.

The bones of the pelvis in the female are flexible and in the male they are not. The most advisable is to take them to the veterinarian so that by laparo-endoscopy they can determine the sex with complete certainty.

COPY

There are actually some conditionals that in the wild determine when these birds begin the reproductive cycle.

- That they find food easily.
- That the weather be favorable.
- May the days be longer.

These conditions appear in <u>spring-summer </u>in many countries, but in Cuba our birds breed practically all year round. Although the reproductive peak is from September to March.

Therefore, you should be concerned about two factors:
Feeding and light. We have already talked about feeding in the previous chapter.

Regarding the light, install one or two cold light bulbs and turn them off

rigorously at 10 pm, they will feed their young until that time, when you turn off the light wait a few seconds and turn it on again seconds later, they will soon learn to find the nest hole at the first warning and retire happy to rest this process of light extension allows to increase the photoperiod.

FECUNDATION

There may be several factors that influence fertilization:

Age of parents, males should be not less than 10 months old and females not less than 12 months old.

Perches in psittaciformes must be well fixed, as their movement may hinder mating, resulting in sterile eggs.

The food must be optimal. They must be healthy.

The illumination period should be sufficient.

LAYING, INCUBATION AND HATCHING.

In Passeriformes and Psittaciformes, approximately 15 days after fertilization, eggs will begin to be laid. Once the first egg is laid, she will continue laying up to a total that varies between 6 and 8 eggs.

During this period the female should not be disturbed, nor should she be cleaning the cage because any manipulation may frighten her and she will stop laying or start to lay her eggs outside the nest.

Some females will start incubation after the first egg is laid but others will start incubation after the second or third egg.

In **passerines**, approximately 15 days after the start of incubation, the first egg will hatch, which will continue daily until the end of the incubation period.

In **Psittaciformes**, between 18 to 21 days after starting to incubate, the hatching of the first egg will occur, which will continue every other day until the end of the incubation period. During this period the female will remain almost all the time in the nest and the male will be in charge of feeding her.

Normally the hatching of eggs will occur daily or every two days, depending on the species: Generally it is not necessary to open the door of the nest to know that the hatchings have occurred as the chirping of the young is characteristic, however it is necessary to do so to monitor that there has not been any death, which is frequent especially when the mother is first-time and this forces us to remove the deceased chick from the nest, as well as to monitor the size of the chicks as between the 5th to 7th day we will have to start banding, which as is understandable will be done daily or every other day depending on the species.

The shells of the hatched eggs will be removed from the nest, as these can affect the new tenants in different ways.

EGG HAULING

Herding consists of transferring an egg from one pair to the nest of another (wet nurse) so that the latter can incubate and raise it in order to save a valuable egg in case the biological mother is unable to do so for any reason.

Generally the female lays after having laid the second or third egg, before that date a cold egg does not necessarily mean that it is not viable.

On the other hand, an incubated egg may not be fertilized, so the temperature of the egg has a relative value.

A fertilized egg <u>after 3 days of incubation</u> shows a lattice of arteries and veins observable through sunlight or a light bulb or at least it is dark or black inside and the shell surface changes becoming smooth.

The wet nurse should be a mother with proven experience as a breeder.

The number of eggs that the wet nurse has to take care of must be taken into account, so that they do not exceed 4.

Egg gathering can be done 5 days before or 5 days after the wet nurse has started laying.

In the **Passeriformes** it is the **<u>Japanese Sparrows</u>** that we use as wet nurses.

BREEDING DEVELOPMENT

In Passeriformes, hatching will occur daily, while in Psittaciformes it will be every other day.

When the birds are born they have no feathers and are pink in color, their beaks are large in proportion to their overall size and their eyes are closed.

At approximately 5 to 7 days, they should be banded even if they have not opened their eyes, which occurs around day 9.

Banding is very useful and is mandatory in certain circumstances.
During the whole time the mother will feed them by regurgitation and they will remain huddled on top of each other in search of warmth.

The newly hatched bird will gain vertiginously in weight. Until about day 23, when it reaches its maximum weight, it is logical to understand that they should not be separated from the parent before that date, since the parent will continue to feed it additionally.

The first thing they do is peek out of the nest hole, then the parents will feed him when he sticks his head out of the hole and in many cases the mother will delay feeding to encourage them to come out of the nest.

Hungry parents finally get it to leave the nest and go for its first walks

Sometimes the chick does not manage to return to the nest on its own, in which case it is advisable to carefully reintroduce it back into the nest in the afternoon.

When **PASERIFORMES** breed in an open nest their characteristics are different.

RINGING

Banding is mandatory in the following circumstances.

To participate in competitions, for marketing and to establish a genetic and veterinary control of the aviary.

The band unequivocally defines the particular bird with a number that is unique to it, as well as identifying the breeder who obtained it and the year of birth.

This procedure is indispensable, even to find out which parent couple it belongs to.

Between 5 to 7 days after birth we should band them, even before they open their eyes, which happens around the ninth day. For the beginner this operation is the most difficult of all the breeding process, the fear of damaging them with the manipulation is unavoidable for some time. Taking them in the hands, holding them in it and handling them creates at the beginning real anxiety in many breeders, only with practice the novice fancier gets over it.

The following system appears in the international bibliography.

Take the young pigeon in the palm of the hand, hold the head between the index and middle fingers, with the big finger and the ring finger hold the leg in such a way that, banding introducing first the 3 biggest fingers and then moving the band towards the fourth smallest finger, recommending banding in the early morning when the young pigeon has not yet fed.

INDEPENDENCE

The chicks will receive food from their parents' mouths even when they are able to eat on their own. That is why it is not advisable to take them away from them too early, because in their company they develop better and if they are weaned too early they will eat less than necessary and will not grow at the normal rate. Only when the parents chase the offspring because they want to nest again is it then necessary to take them out of the cage as a matter of urgency.

The following precautions shall be taken during independence:

- Never do independence in the afternoon.
- On the contrary, provide clean water and food early in t h e morning and when you see that it has eaten, take it out.
- In the new cage put water and food.
- Do not put the chicks directly into a very large cage.
- Put it in a cage similar to the one it was in.
- Only when he is there for a week will he then make the transfer to the flight cage.
- A good procedure is to place the weaned chicks together in that cage and transfer them all together to the cage.

GENEALOGICAL CARDS.

These cards allow the control of many of the factors of great importance in poultry breeding, such as:

Color, ring number, genetic formula and carrier characteristics of the sire, dam and grandparents, as well as the date of birth of each of the offspring.

Example: Cage no.   23

	Female		Male	
Color				
Ring				
Carrier				
	Father of the female	Mother of female	Father of the male	Mother of the male
Color				
Ring				
Carrier				
1 Settling	2 Settling		3 Settling	4 Setting	5 Setting
Date and Ring No.	Date and Ring No.		Date and Ring No.	Date and Ring No.	Date and Ring No.

THE PSITTACIFORMES

Psittaciformes are characterized by four toes, two forward and two backward, and their beaks are curved and strong.

CLASSIFICATION

PERIQUITES

Corrugated

Opalinos

Black-eyed yellows

Whites with black eyes

Lutinos

Albinos

Flavos

Harlequins

Spangles

Lace wings

Clear body

Saddleback

Moñudos

ROSEICOLLIS

Green Series

  Lutinos

    Australian yellow

    American Yellow

   Japanese yellow

  Dorados

  Golden cherry

  Pardinos

From the blue series

  Blue

  Cherry Silver

  Silver

  Albinos

  Creminos

  Pardinos

Harlequins

Canelas

White mask

Orange mask

Violets

Cakes

FISHERIS

Green Series

Blue series

Lutinos and Albinos

Diluted

Harlequins

cinnamon

Silver

gilded

PERSONATAS

Green and blue series

Diluted

Harlequins

Canelas

Albinos and Lutinos

Gold and silver

NYMPHS OR CAROLINAS

Gray

Targets

Canelas

Tan and gray harlequin

Pearlized

PARAKEETS VARIETIES

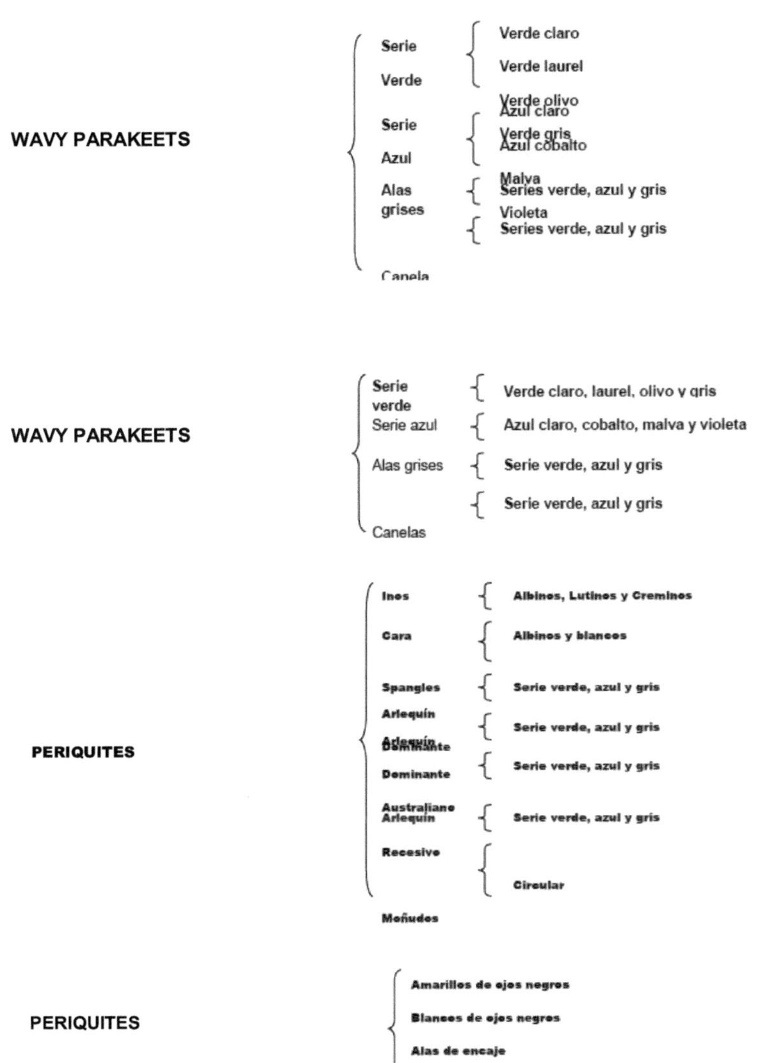

PERIQUITES

Budgerigars originating in Australia were discovered in 1789.

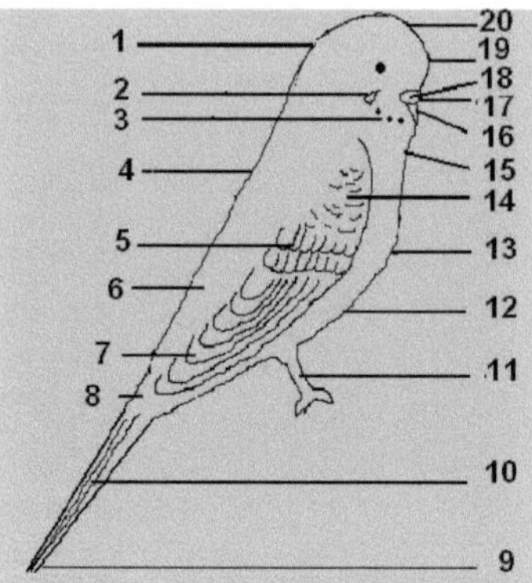

1. nape, 2. whisker, 3. pearls, 4. back, 5. secondary remnants, 6. rump, 7. shirts. primary, 8. caudal coverts, 9. rectrices, 10. rudders, 11. legs, 12. abdomen, 13. chest14. wing coverts, 15. neck, 16 beak, 17. wax, 18. nostril, 19. forehead, 20 crown.

COLOR IN THE PARAKEET

In some texts it appears:

- **Green dominates all other colors.**
- **Yellow dominates over white.**
- **Blue dominates over yellow and white**
- **Gray and violet dominate over blue, yellow and white.**
- **There is co-dominance between gray and green.**

In reality, although some speak of dominance, **what really happens here is an effect of absence and / or presence,** additionally the genes determining the colors are not linked, because they are located in pairs of different chromosomes, but strictly speaking, when we involve sex-linked colors, then there are linked genes and also in the case of lipochromes and the dark factor, because they are in the same pair of chromosomes, the formula 2 raised to "n" is not fulfilled since they are not distributed randomly as a result of gametogenesis.

COLOR SHADES

Tinting.

1- The dark factor.

2- Australian dominant gray.

3- Violet.

THE DARK FACTOR.

Normally the recessive characteristics need to be present in homozygous form to be expressed phenotypically, since in the heterozygous form they remain as carriers. However, in intermediate inheritance, heterozygosity is expressed phenotypically although with less intensity, an example of this is the dark factor.

We have already seen the genesis of the appearance of the colors: white, yellow, blue, and green, (Page: 3) but that does not explain the existence of the shades within these colors because as we know they exist within the green series:

Olive green, laurel green and light green. In the yellows we find:

Light yellows, medium yellows and dark yellows
As well as within the blue series there are:

Light blue, cobalt blue and mauve blue

These three gradations are differentiated by their greater or lesser darkness and are due to a gene called dark, such a mutated gene appeared between 1915 and 1920 and superimposes its pigmenting action on the normal gene.

If we take into account that the dark gene can act in single or double doses, it is easy to understand the following scheme:

Green color with no dark gene............ Light green Green color with

one dark gene...... Laurel green Green color with two dark genes...

Olive green Similarly in the blue series we see:

Blue color with no dark gene.   Light blue Blue color with

one dark gene...... Cobalt blue Blue color with two dark genes. . .

Mauve blue **THE GRAY FACTOR**

The gray factor masks the other colors, giving in the green series a <u>dull, dull olive tone </u>(the intensity of which varies according to the amount of dark factor the bird possesses, which is why the untrained, e.g., think they have an olive green and are actually looking at a gray green). In the blue series it leads to a <u>light gray </u>tone<u>.</u> This factor acts similarly in all other varieties.

THE VIOLET FACTOR

The violet factor alters the color intensity in the blue series so that light blue with violet factor looks cobalt. When cobalt and violet are combined, a beautiful violet color results,

Some authors point out that the violet factor only appears phenotypically when a specimen possesses, in addition to the blue factor, the dark factor and the single or double violet factor. Although it may seem strange, it is possible to have entire series of green birds with the violet factor, being then known as violet green.

WAVY PARAKEETS.

The wavy parakeets have the base of the head, nape and neck with black undulations on a green background, the mask is yellow and on it three black polka dots symmetrically distributed on each side. The green wavy parakeet is the one that exists in free form in nature and which we identify as "**wild**". Within the green series we find **light green, laurel green and olive green**, but there are also other mutations such as gray, gray-green and violet. And in the blue series, light blue, cobalt and dark blue, wavy with gray or cinnamon wings may appear.

GREEN SERIES CORRUGATED

We have already seen that within the green series there are three color gradations. Among them, the laurel green is the most interesting from the reproductive point of view, since the coupling of two laurel greens offer us a brood where we can find: **light green, laurel green and olive green**. The difference between olive green and gray green is that <u>in gray green the whiskers are gray</u> and as the gray infiltrates into the green it dulls the brightness resembling the olive color. In olive green the whiskers are violet.

The recessive form of light green is yellow, where three gradations appear.

Characteristics: Mask: yellow

Moles: three on each side of the ruff, black in color, symmetrically distributed and approximately the size of the eye.

Whiskers: violet and Eyes: dark with visible iris.

The basic color, depending on the case, can be light green, laurel green, or olive green. It is distributed pure and uniformly on the chest and belly covering the neck, back, obispillary and wings, serving as a base to support the melanic design.

Melanins: black (eumelanin) distributed on the head, neck and wings.

T-shirts: greenish black. Rudders: dark blue and legs: bluish gray.

BLUE SERIES CORRUGATED

Here again the most interesting thing is to couple a cobalt blue with another cobalt blue where we will obtain in the brood both light blue, cobalt blue and mauve, although as in the previous case, in different proportions.

The recessive form of blue is white. Characteristics: Mask: white

The basic color, depending on the case, may be light blue, cobalt blue

or mauve.

The rest of the characteristics coincide with the description above.

WAVY GRAY AND VIOLET

Both factors are dominant, so we cannot assume that a specimen that is not gray or violet can be a carrier of this characteristic.

Regarding the violet factor, there is only a very specific group "the cobalt violets" that clearly show a violet tone in their plumage, all the other specimens present variations in their coloring, but they do not appear in this color to our sight.

The manifest form of the violet color occurs when the blue factor, the single or double violet factor and the single or double dark factor coincide in the same specimen.

Single dark factor = cobalt = visible violet Double dark factor = mauve

= non-visible violet.

GRAY-WINGED RIPPLES

They can appear in all shades: Green, Gray Green, Gray and Blue.

In these cases the basic color is diluted by the influence of gray, distributed in the neck, chest, belly, back, obispillo and wings, serving as a base to support the melanic design that will be light gray.

WAVY CINNAMON WINGS

They can appear in all shades: Green, Gray Green, Gray and Blue.

In these cases the basic color is diluted by the influence of the cinnamon, (pheomelanin) being distributed in the neck, chest, belly, back, obispillo and wings, serving as a base to support the melanic design that will be of cinnamon color.

OPAL PARAKEETS

The opalines, the feathers of the wings have the color of the body, the feathers are dark <u>with wider edges than in the wavy ones</u> and can be greenish or blue depending on the series.

GREEN AND BLUE SERIES OPALINES

Opalines have the base of the head, nape and neck without

undulations. The feathers of the wings have the color of the body.

The ruffled feathers are dark with greenish or blue edges depending on the series and wider than in the wavy ones.

On their wings there is a 1 to 2 cm. interruption in yellow or white, depending on the color series, which is usually called "**<u>opal shield</u>**". Another distinctive feature of the opaline is that it marks its melanins at the shoulders **<u>defining a perfect "v"</u>** that distinguishes them.

The mask in the form of a hood, as clean as possible of melanin, extending over the head, neck, shoulders and neck.

The melanin is evenly distributed on the wings **<u>in the form of teardrops</u>**. They can appear in all shades: Green, Gray Green, Gray and Blue.

OPALINE GRAY WINGS

In these cases the color is diluted by the influence of the gray, it is distributed on the back marking a "v" obispillo and the lower part of the ruff to behind the legs.

The gray melanin is evenly distributed on the wings in the form of teardrops.

They can appear in all shades, Green, Gray Green, Gray and Blue.

OPALINE CINNAMON WINGS

In these cases the basic color is diluted by the influence of the cinnamon marking a "v", obispillo and the lower part of the ruff to behind the legs, with melanins and tan remiges, a combination that appears by Crossing over in certain mating of cinnamon and opal specimens, which are two sex-linked recessive mutations that travel on the same chromosome.

They can appear in all shades, Green, Gray Green, Gray and Blue.

YELLOW AND WHITE WITH BLACK EYES

Yellows are yellow all over the body, except for the rudders, which are white and yellow.

The mask will be yellow or white depending on the case. They lack

moles.

White mustaches.

The eyes are black with no visible iris.

In the whites the color is white all over the body and some yellow may appear in the remiges. Pink legs and wax: lilac in males.

INO FACTOR, LUTINOS

Yellow mask, same as the entire body

They may or may not have moles; if they do, they are very light gray

in color White whiskers

Yellow color all over the body except for the remembranes and rudders, which are white, **the nasal wax is pinkish mauve in males**.

The legs are flesh colored or pink. The eyes are red.

INO FACTOR, ALBINOS

White mask, same as the whole body.

They lack moles and well marked white whiskers.

White color all over the body, **pinkish mauve nasal wax in males**.

The legs are flesh colored or pink and the eyes are red.

FLAVOS

Yellow mask in the green, green-gray and white mask in the blue and gray series.

Red eyes and dilute violet whiskers.

The rest of the features such as polka dots, remiges and rudders vary to a lighter cinnamon almost brownish.

ARLEQUINS.

The harlequin factor has the characteristic of not allowing the manifestation of melanin in certain areas of the body. Fundamentally in the ventral area, back of the head, the primary feathers of the wings and rudders of the tail.

There are three varieties of harlequins: Australian dominant, Dutch

dominant and recessive Danish.

AUSTRALIAN DOMINANT HARLEQUIN.

They have a color band that divides the base color into two parts.

They have a spot on the head, light-colored remiges and rudders and a well-defined band in the central part of the chest, yellow or white depending on the series, which divides the belly into two parts.

His whiskers are violet. And their legs are dark.

Very good quality specimens show the complete set of pearls on the throat, as well as impeccably drawn ear spots.

They can appear in all series: green, green-gray, blue and gray.

The difference between the two green series is that in the gray green the whiskers are gray and as the gray infiltrates the green it dulls the brilliance.

DUTCH OR CONTINENTAL HARLEQUIN

It is distinguished by a light spot (aureole) on the upper part of the nape of the neck of yellow or white color depending on the color series to which it belongs and by the light tone of its remiges and rudders. With white whiskers and pink legs.

The body divided more or less in half into green and yellow or blue and white depending on the green or blue series.

RECESSIVE HARLEQUIN DANISH

They have a yellow or white mask and moustaches in violet and white or gray and white, depending on the series to which they correspond.

Black eyes without white rim, (**without visible iris) fundamental characteristic**. In this variety the wax of the male is pink and in the female it is brown.

To improve their technical marks, a wavy first generation carrier should be crossed with **a** recessive recessive manifest Dane.

It has been observed that first generation carriers obtained from the cross of harlequin with a wavy gave better results than carriers from many generations of crosses with harlequins.

YELLOW MASK

The yellow mask has two mutants, I and II, which we will see later.

It is usual to mate yellow-faced parakeets with wavy-faced parakeets of the blue series, because in the blue series the yellow mask makes a beautiful contrast.

A very beautiful specimen *THE ANTIQUELY CALLED RAINBOW* is nothing more than an opaline, blue, gray wings, yellow mask.

All budgerigars of the green series can **potentially** be carriers of yellow face, but as it is to be expected from their external appearance, they will not differ much from the normal green ones.

Although in carrier greens, the yellow becomes more intense in the mask.

Combination of varieties:

Blue spangle yellow mask.

AGAPORNIS ROSEICOLLIS

Here are some descriptions that may be useful but are not technical standards and cannot be interpreted as such. They are intended to provide the beginner with general information for the identification and classification of the different varieties.

DESCRIPTIONS GREEN SERIES

The roseicollis of the green series can be of three types, light green, laurel green and olive green.

The crossing of two laurel greens has the special characteristic that the three shades of green can be obtained from this cross: light, laurel and olive.

BLUE SERIES

In this series we will also find three shades: light blue, cobalt blue and mauve.

Here we must point out that mauve blue will actually have a color more similar to gray than to mauve. This is clearly observed when we mate two cobalt blues and in the offspring we obtain light blues, cobalt blues and mauves, making it clear that mauve is the shade that responds to the possession of two dark factors in the blue line.

VIOLET

This mutation is dominant and can appear in single or double factor.

It reflects the shortest wavelength color available which is additionally influenced by the dark factors that can be added to it.

In this case it is important to distinguish the double violet factor specimen from the violet specimen with double dark factor, which is not the same thing.

This is a recent mutation

It is not advisable to pair it with the green series.

AGAPORNIS PERSONATA

They live in Tanzania and have a periophthalmic ring.

They carry nesting material in their beaks. They have no sexual dimorphism, so both sexes are the same, with a black mask covering their heads.

The throat and chest are yellow and the rest of the body is green, the rump is blue.

Nests should be somewhat larger than those used for other species.

It is advisable to breed them in pairs, although some have used the colony system.

Among its varieties we have: Light blue, cobalt, mauve, light green, laurel, olive, diluted, Lutinos, harlequins, albinos, gold and silver.

AGAPORNIS FISHERI

It is one of the three most popular species today. They live in Tanzania and nest in colonies in the wild, but breed very well in captivity, especially in colonies, which is logical considering their natural habits. They carry nesting material in their beaks. They have no sexual dimorphism. The forehead is reddish orange as are the cheeks and throat. The neck is yellow while the back of the head is olive green, the rump is bright blue and they have a periophthalmic ring.

Among its varieties we have: Green and blue series Harlequins,

Canelas, Silvers, Dilutes. Lutinos, Albinos and Dorados.

TARANTA OR BLACK-WINGED LOVEBIRD

They live in Ethiopia, it is easy to recognize the female from the male because the male is green with a red stripe on the forehead that extends to form a circle around the eye, which does not occur in the female (sexual dimorphism). The remiges of the male are black, while in the female they are brownish black. The beak is red in both.

GRAY-HEADED LOVEBIRD

They live in Madagascar, and is one of the smallest 13 cm. It presents sexual dimorphism.

In the male, the head, neck and upper chest are light gray but in the female these parts are light greenish yellow.

In the wild they live in large groups but do not nest in colonies.

Breeding presents serious difficulties.

BLACK-NECKED AGAPORNIS SWINDERNIANA

(Known as Swindern)

They live in central Africa, staying most of the time at the top of the trees, making them difficult to see. It is the only species that has a black beak with a black band on the back of the neck and a brownish yellow band immediately below it. To our knowledge there are no specimens in captivity.

AGAPORNIS PULLARIA OR RED-FACED LOVEBIRD

They live in central Africa, it presents sexual dimorphism, although it is not easy to delimit the difference, the male has a red face that extends from the forehead and reaches through the center of the eye to the front of the neck, the beak is also red, the female has the same characteristics but more orange than red that merges with an almost yellowish plumage, the color under the wing coverts is green instead of black as in the male. It is a very shy bird, and it is very difficult to breed in captivity because they build the nest in earthy mounds where they dig a kind of burrow where they lay their eggs.

AGAPORNIS NIGRIGENIS OR BLACK-CHEEKED LOVEBIRD

They live in the sub - central part of Africa, the color of the head is brown with black cheeks and face but the upper part of the chest is orange red, has periophthalmic ring. and do not have sexual dimorphism, the rump is green.

Paradoxically, although they breed easily in guardianship, it is a species threatened with extinction. In the past they were imported in large quantities but undesirable crosses have made it difficult to find pure specimens due to genetic erosion.

AGAPORNIS LILIANAE

They live in Zambia, Tanzania and Mozambique, they have a red bill, the crown and forehead are orange which gradually dilutes to become salmon red around the throat and cheeks, their resemblance to the fisheris is great, but the Fisheris has blue upper tail feathers while the rump of the Lilianae is green. The lutino mutation of the Lilianae has a red head while the body is bright yellow with white remiges. Because they are not very aggressive, it has been possible to breed them in colonies, even though many advise breeding them in pairs.

CACATILLOS

Among the Cacatillos we will find:

gray	Cremino
pearl gray	lacewing
harlequin gray	pardino
harlequin pearl gray	pastel gray
white-masked gray	harlequin pastel gray
gray pearl with white mask	pastel pearl gray
gray harlequin with white mask	harlequin pearl gray pastel
harlequin pearly gray pearl white mask	gray mask white pastel
albino	harlequin gray mask white pastel
lutino	pearl gray pastel white mask
pearl lutino	pearl gray harlequin white pastel mask
yellow	cinnamon pastel
cinnamon	harlequin cinnamon pastel
pearl cinnamon	cinnamon pearl pastel
cinnamon harlequin	harlequin cinnamon pearlized pastel
cinnamon pearl harlequin	cinnamon white pastel mask
white mask cinnamon	cinnamon harlequin white pastel mask
cinnamon pearl white mask	cinnamon pearl white pastel mask
white mask cinnamon harlequin	harlequin cinnamon pearl mask white cake
mascara cinnamon pearl harlequin white	silver flavo
silver	olive tree
silver harlequin	pastel side
white-masked silver	white with black eyes

PASSERINES

It is appropriate to know that all the species we will see below: Canaries, Japanese Sparrow, Java Sparrow, Gould's Diamond, Mandarin Diamond, Long-tailed Diamond and Sparrow are Passerines which have four toes, three of them are directed forward and only one backward, while their beaks are straight, generally short and do not have great strength in them.

In addition to lipochromes and melanins, carotenes may also play a role in the color of passerines, the constitution of which mainly involves carbohydrates.

CANARIES CLASSIFICATION

IN COLORED CANARIES ACCORDING TO THEIR MELANIC CHARACTERISTICS

Lipochromics	**Melanics**
White	Black Brown, Cinnamon and Agate
Dominant white	Elizabethans
Yellow	Cakes
Red	Gray wings
Ivory red	Opalinos
Albinos	Satine and Eumo
Lutinos and Rubinos	Onyx and Topaz

HISTORY

The canaries took the name of the islands where they were discovered "Canary Islands" and according to English documents dating from the sixteenth century was Queen Elizabeth I in 1580 who began breeding from a pair that she received as a gift and that with great care and care managed to reproduce, which eventually became a royal gift that the sovereign offered to her nobles as a sign of special deference.

Other documents indicate that breeding began in some Spanish monasteries where the monks, encouraged by the high price at which they could be sold, specialized in it, but only marketed the males, keeping the females in their possession in order to monopolize the trade, although the truth is that Queen Isabella somehow managed to obtain a couple with which she began their reproduction.

But it was not until 1670 that the mutation to the much-loved yellow color occurred and quickly spread to the other continents. Soon

new mutations appeared and today there are more than four hundred chromatic varieties

THE COLOR OF THE PLUMAGE

LIPOCHROMES AND MELANINS

The color of the plumage in canaries is given by chemical, physical and biological factors. The chemical factors depend on biochemical structures called pigments. In general pigments can be of two types

DESCRIPTION

These descriptions are not technical norms nor can they be interpreted as such, they are only intended to facilitate the identification of the different varieties by the beginner.

LIPOCHROMIC CANARIES

WHITE CANARY

Here the sedimentation of   all
pigments is totally prevented, giving these canaries a clear white color, with no traces of yellow.

(Total inhibition of lipochrome) WHITE CANARY DOMINANT

In these specimens the sedimentation of all pigments in their plumage is partially prevented, so their appearance is white, leaving only traces of yellow on shoulders, shoulders and rudders.

(Partial inhibition of lipochrome deposition in the plumage). YELLOW

CANARY

It has a uniform yellow color.

Lemon yellow is preferred over golden yellow, orange is not accepted.

CANARY RED

It arises from the hybridization with the Venezuelan Cardenalito from which it inherits the ability to assimilate red pigments giving a bright and uniform red coloration.

IVORY RED CANARY

When affected by the ivory mutation that softens and softens its color, it takes on a pinkish hue.

IVORY YELLOW CANARY

It has a light yellow color similar to straw color, as a consequence of the influence of the ivory mutation.

RED-EYED LIPOCHROMIC CANARIES

All lipochromic canaries "can" present the particularity of having red eyes, which can be due to two different mutations. INO AND SATINE.

In competitions, red satin eyes are preferred.

MELANIC CANARIES

BLACK BROWN

It has all three melanins oxidized.

The melanin structure shall be completely jet black, the dorsal pattern shall be broad, strong, uninterrupted and without diluted areas.

The black eumelanin is oxidized to the maximum, reaching the very edge of the rims and rudders. With completely black legs, toes, nails and beak.

Depending on the lipochrome has different names to better define the variety, so when the background is:

White dominant, we call it: blue dominant It is the brown black with white dominant background, offering a lead gray shade.

Recessive white or simply "white" (we call it: blue). It is the brown black with recessive white background, with total absence of the lipochrome, offering a tone similar to silver.

Yellow, (we call it: Green, It is the brown black with the yellow background that gives us the specimen that we call green by the superimposition of the yellow. The green color should be clearly seen on the chest, belly and between the melanic pattern.

Red, (we call it: Copper). It is the brown black on a red background that makes us appreciate a color similar to copper, its colors should be uniform, bright and harmonious, without presenting areas of different shades.

<u>Ivory Yellow,</u> (we call it: Ivory Green). It is the brownish black with yellow background and the ivory mutation added on, it offers a luminous, harmonic and soft light green shade.

<u>Ivory red,</u> (we call it: Copper marfi) I. It is the brownish black with red background and the ivory mutation added on, offering a coloration with a bright dark pink shade.

CANARY CANARY

It is a sex-linked recessive mutation that has acted on the black eumelanin modifying it into brown eumelanin.

It has the same melanic pattern as the brownish black but with a brownish color.

It has brown eumelanin and pheomelanin oxidized but black eumelanin reduced to brown. <u>Depending on the background they appear:</u>

<u>Cinnamon yellow,</u> very warm ocher color of great beauty for its

softness. <u>Cinnamon red,</u> provides a dark red hue to the eye.

<u>White cinnamon,</u> is the dominant silver with partial absence of the lipochrome, the scattered distribution of brown eumelanin should cover it completely, the recessive silver is similar to the previous one but the absence of the lipochrome is total so it does not present any type of incrustation.

<u>Ivory cinnamon,</u> the superimposition of the dispersed brown eumelanin on the yellow diluted by the effect of the ivory factor gives us a much lighter shade than the yellow cinnamon, resulting in a silkier, softer, bushier and more compact plumage.

<u>Ivory red cinnamon</u> is similar to red cinnamon but due to the action of the ivory it has a much lighter, pinkish background color.

It is one of the most robust and easy to breed canaries.

Highly recommended to start breeding by amateurs who are taking their first steps. It is considered the first mutation of the original or wild green canary.

AGATE CANARY

It is a sex-linked recessive mutation producing a melanic reduction called dilution and is named for its agate-like color. It is the diluted brownish black. All three melanins are diluted.

It is the diluted form of green, i.e. it is a diluted green.

The mutation known as Agate affects the black eumelanin in the sense of concentrating it towards the center of the feather, leaving the edges almost totally depigmented and the pheomelanin reduced to the maximum.

The mantle of the Agate is presented in the form of short and narrow strokes. Agates have eyebrows with almost clear lipochrome and whiskers are very marked and perfectly contrasted, the melanins of the head are perfectly striated and marked, the whiskers stand out sharply.

The whole melanic pattern should be dark gray almost black, starting behind the beak. The edge of the feathers is pearl gray.

Good Agate should have a total absence of Pheomelanins.

Depending on the background they appear: Agate yellow, red, white,

ivory and ivory red. ISABELA CANARIES

It is the product of the interaction between the mutation that transforms the black eumelanin into brown and the dilution existing in the agate. It is the diluted form of Cinnamon, i.e. it is simply a diluted Cinnamon. These specimens appeared by "crossingover" between Cinnamon and Agate. It has reduced pheomelanin.

The beak, legs, toes and nails should be flesh colored. EUMO

CANARY

The Eumo mutation is autosomal recessive, it centralizes and reduces melanins both longitudinally and transversely.

These specimens have a striated design and the space between the striae is luminous, the melanin is in the axis of the feathers.

They have an almost total absence of pheomelanin with a centralized melanic pattern.

They have red eyes.

Legs, nails and beak are light colored.

Appearance: Eumo brown black, Eumo agate, Eumo cinnamon,

Eumo isabella. OPAL CANARIOS

It is an autosomal recessive mutation that acts on melanins by reducing and inverting them, producing in brownish blacks, agates and browns a bluish-gray appearance similar to the opal stone that gives it its name.

This mutation transforms the black eumelanin into a bluish-gray shade that is well defined in the specimens of the Brown Black and Agate series and with less definition in the Canelas and Isabelas. Since Opal is a mutation that affects mainly the structure of the feather, the horny parts of each type will remain identical to those of their respective classical types.

Appearance:

Black Opal brown. Cinnamon Opal.

Agate Opal. Opal Isabela.

INOS OR UGLY CANARIES

We call "FEOS" the melanic canaries affected by the "INO" mutation of autosomal recessive transmission that inhibits the two Eumelanins, the inhibition of the black eumelanin, includes those of the beak and legs, as well as the eyes, but respects the pheomelanin, giving it a greater melanic charge.

These specimens present sexual dimorphism, so the male is perfectly distinguishable from the female, which has a greater amount of

Feomelaninas. In addition, the male presents a facial mask. Beak and legs appear flesh-colored in them, although there are variations depending on the type. But the eyes are red. They appear: Rusty ruby (the Black Cinnamon Rubino and the Cinnamon Rubino) and dilute ruby (the Agate Rubino and the Isabella Rubino).

According to the background: (Red = Ugly Rubino) (Yellow = Ugly Lutino) (White = Ugly Albino dominant) (Ivory Lutino = Ugly Lutino ivory) = (Ugly Albino. White).

PASTEL CANARIES

It is a sex-linked recessive mutation that almost totally reduces pheomelanins while respecting black eumelanin.

The effects of the PASTEL mutation can be defined as a reduction of pheomelanin, dilution of black eumelanin and dispersion of brown eumelanin.

The pastel brown black should have a more diluted melanic pattern than the classic brown blacks and its shade goes from black to grayish.

Pastel Cinnamon, the action that the pastel mutation exerts on the classic cinnamon canary results in a notable reduction of the pheomelanic structure and the dispersion of the eumelanin, so that the eumelanic pattern is considerably reduced in size. The general appearance is of a dark brown tone with a small pattern on the back, head and flanks.

Pastel Agate, when the pastel mutation appears on an Agate canary, the weak pheomelanic structure of the latter is sensibly reduced by the pastel factor until it almost disappears in some specimens that are considered optimal.

Pastel Isabela, the reduction of the phaeomelanic structure and the total dispersion of the eumelanic pattern, all due to the influence of the pastel mutation, make the pastel Isabela totally lack a dorsal pattern, either on the head or on the flanks, leaving both melanic structures amalgamated, presenting only a light brown mantle that surrounds it.

SATIN CANARY

It is a sex-linked recessive mutation that completely eliminates the black eumelanin and pheomelanin, respecting the brown eumelanin, with red eyes. Its melanic pattern is identical to that of the Isabella but without a trace of pheomelanin, so that between the gaps in the pattern, the lipochrome of the background emerges completely clean.

Depending on the background they appear: Satine yellow, red, white, ivory and ivory red.
TOPAZ CANARY

It is an autosomal recessive mutation on melanins and codominant with respect to the Rubino mutation. It is characterized by the concentration of melanins around the medullary center of the feathers and the luminosity of the background lipochrome that appears is due to the absence of Feomelanins.

This mutation reduces black eumelanin and pheomelanin and also acts on the beak, legs and nails, giving them a lighter or lighter shade of brown.

Appearance: Brown black, Topaz, Topaz agate. CANARY ONYX

In this autosomal recessive mutation occurs the opposite of what happens in the other mutations seen previously that dilute the melanins, here the mutation darkens them all, giving very different shades according to the phenotype present, becoming evident a concentration of Eumelanins from the lower part of the neck to the head of the specimen.

SINGING CANARIES

There are three types of singing canaries: Harzer Soller, Malinois and Timbrado. These three species differ mainly in their different singing qualities.

The particularity of these canaries is that they are not required to have beauty of form, colors, size, drawing, etc. Because their only merit resides in their song, in their melody, in the notes they emit, which must conform to a correct standard.

In the exhibitions, their judging is done in an environment outside the vortex of the exhibition, this is because every sound or noise and especially the singing of the other birds exposed would impair the sharpness and fluctuation of the notes and the clarity of the timbres that are required in the competition to these birds.

The line of selection followed in the Harzer has made possible canary characteristics that cannot be compared to any other canary.

No song canary requires special care in terms of feeding or breeding that differentiates it from colored canaries, but only expert breeders are able to rationalize and program the laying in such a way that lines of good singers are established.

It is difficult and complex to state here the characteristics that they are supposed to possess, but to get an idea, think for example of the sounds and noises that the males should emit when in the presence of the judges and which should resemble deep rolls, chained rolls, water noises, deep ringing sounds, flute notes, cooing sounds, clucking sounds, etc.

ACQUISITION

When purchasing his first pair of singing canaries, the novice fancier should <u>always</u> seek the advice of an expert breeder, as only a trained ear is capable of assessing the quality of the song.

TRAINING

The first thing is to have a suitable place for training, nothing that distracts the canary, away from the females and other birds. The place should have artificial light.

The song canary is trained in wooden cabinets 22 inches long, 20 inches high and 7 inches wide, will have two doors and will be divided into 4 equal departments of 11 inches long, 8.5 inches high and 6.5 inches wide, divided by a cross-shaped board. The small cages inside the cabinet are 8.5 inches long, 8 inches high and 6 inches wide.

Singing canaries begin their apprenticeship after molting, which takes place at 7 months or more after birth.

Training begins in September to compete in December.

The placement of the males will be next to each other if they are siblings and then the most closely related: This results in the training being dominated by the hereditary characteristics of the family.

In this way, the sets are formed for future appraisals and exhibitions.

The chicks, already in the cages with water and food, are kept there for 3 or 4 days, always with artificial light, after 8 or 10 days they can be placed in the cabinets.

The breeders sit daily in front of their favorites and observe the progress of the singing, analyzing the hours of the day in which the chicks are more prone to sing and eliminating those that in their opinion do not fulfill the required singing characteristics, the best ones are chosen and the darkening stage begins with them, which should be progressive. This operation is done after 8 or 10 days when the canaries already know where the water and food are.

The cages are first darkened by closing the cabinet doors and after approximately 3 weeks, the room is darkened for two hours in the morning and two hours in the afternoon.

The 4 basic tasks to be issued are: Hahlrolle, Knorre, Hahl Kingler and Pfiefe.

It must be kept in mind that too intense darkening inhibits the development of the song, the right degree is when the chorus of the chicks emit their song to the fullest when the singing cabinet is opened.

After this stage, they are placed on a table, one cage on top of the other and are listened to 3 times a day, for 15 to 20 minutes, always with artificial light. Classifying them and placing them in cages numbered 1, 2, 3 and 4.

Finally, put in cage 1 (head canary, the best Hahlrolle and Pfiefe), and in cage 4 (table canary, the best Knorre, bass)

EXOTIC

Of the great variety of those considered exotic in our environment, we will only detail some of the most frequently started by beginners.

Only on this first page has mention been made of the common name (to help you identify which one they are talking about when someone uses that name).
But from now on we will only use the official nomenclature so that you can adapt to it.

Official Name	Scientific name	Common name
Japan Sparrow	*Lonchura striata*	isabelitas
Mandarin diamond	*Poephila guttata*	zebritas
Gouldian diamonds	*Poephila gouldiae*	lady gould
Bibs	Long tail diamonds	bibs
Paddas	Java sparrow	Hungarians
Sparrow	Emblema guttata	gutatas

But there are other exotic species such as:

strildas	frugivores	american avifauna
ammunition	euplectes	hybrids
grenadines	house sparrow	other diamonds
erythruras	wild avifauna	other exotics

GENERAL CHARACTERISTICS JAPANESE SPARROW

(ISABELITAS)

In Cuba they are known as Isabelitas, but in many books they appear as "Capuchino Japonés". They are small birds, these birds present the colors carmelite, white and mottled. Originally, there was only the dark brown variety and the light brown spotted one. Currently there are a number of mutations such as the dark brown with white belly, yellow, white and reddish.

JAVA SPARROW (HUNGARIAN)

Although in our country they are known as Hungarians, in many others they are called Java Sparrow. Their average size is 14 cm. Their beak is somewhat curved and thick but thin at the end and red to pinkish in color. Its body is steely gray. The head on the dorsal side is black. The cheeks are white. The ventral part is carmelite to light violet and becomes almost white in the anal region. The eyes are dark with a reddish ring around them. The tail is black and the legs are pink.

GOULD'S DIAMOND (LADY GOULD)

Gould's Diamonds are small birds with straight bills, in them nature overflowed coloration and beauty, when observed they give the impression of having been painted by hand because of their well defined colors. They are named after the color of the head, breast and back.

LONG-TAILED DIAMONDS (BIBS)

Although in Cuba they are known as Baberos, in many books they appear as Diamantes de cola larga (Long-tailed Diamonds). Their characteristics are: Dark steely gray dorsal plumage, with light gray belly. The beak is thin, elongated and red. The neck has a black bib on its ventral part (from which it gets its name). The tail is black, ending in a long central feather.

MANDARIN DIAMOND (ZEBRITAS)

Although in Cuba they are known as Cebritas, the POEPHILA GUTTATA is mentioned in other books as Diamante manchado or DIAMANTE MANDARIN. The beak is orange-red in color. The head, in the males has a small black demarcated stripe, the cheeks of orange color, the female has only the small black stripe. It has been possible to breed the white, silver, marbled, black-breasted and other mutations.

SPARROW (GUTTATAS)

The Emblema Guttata in our environment has been called simply Guttatas but this is a bad habit because there are other Guttatas such as the Pheophila Guttata (known in Cuba as Cebrita) and this tends to confuse. In other latitudes it is called Spotted Diamond or also Fire-tailed Diamond. Its size is around 12 cm. The head is gray and the beak is red, it has a black band from the eyes to the beak. The eyes are black with an orange-red ring around them. The gray neck is separated from the white ventral part by a black band that extends to the tail and is speckled with white. The back may be light crimson and the tail is black.

PECULIARITIES OF SOME EXOTIC JAPANESE SPARROWS

(*Lonchura striata* doméstica) It belongs to the Estrildidae family.

In Cuba they are known as Isabelitas, one of the oldest birds from Japan. They are small birds, approximately 11cm. in length and 17 g. in weight, these birds present carmelite, white and mottled colors. Their nests and other characteristics are similar with all the Passeriformes, distinguished for being very good breeders for that reason <u>they are used very frequently as NODRIZAS</u>, generally 3 pairs of them for each pair of another exotic.

They should not breed more than 4 times a year, leaving them to rest the rest of the time.

They lay an average of 5 to 7 eggs on successive days and in the early hours, incubation lasts approximately 14 days where the parents alternate during the day and together during the night. At 20 days the chicks leave the nest, although they are fed by the parents for another 15 days. After that date, they must be separated because they tend to sleep next to their parents, thus hindering a new nesting. Females are ready to breed at 6 months of age.

They clean themselves with their beaks, so they must have a candle (drinking fountain) to drink water and prevent it from getting dirty.

Japan Sparrows among themselves are so sociable that they will all sleep together in the same nest and get along well with other small birds.

However, as they become bellicose at the slightest aggression, they must be kept in individual cages.

Originally, there was only the dark brown variety and the light brown spotted variety. Currently there are a number of mutations such as the dark brown with white belly, yellow, white and reddish. The rest of its characteristics are similar to those seen for all passerines.

JAVA SPARROW (PADDA)

It belongs to the Estrildidae family.

Although in our country they are known as Hungarians, their name is Java Sparrow. It is native to Indonesia.

Its average size is 14 cm. and it weighs approximately 30 g. Its beak is somewhat curved and thick, but thin at the end and red to pink in color. Its body is steely gray in color. The cheeks are white. The ventral part is carmelite to light violet and becomes almost white in the anal region.

Dark eyes with a reddish ring around them.

The tail is black and the legs are pink. Among its mutations are: The white. The mottled and the carmelite among others.

Sexual differentiation is difficult, although in the male the shades are stronger.

The female lays between 4 and 6 eggs, incubation lasts approximately 14 days and after 4 weeks they leave the nest, but continue to be fed for 2 more weeks by their parents, at which time they should be weaned.

The rest of its characteristics are similar to those seen for all passerines.

GOULDIAN DIAMOND

Gould's Diamonds, native to Northern Australia, are small and fragile birds that leave us stunned by their wonderful beauty, their colors are so varied, even and delimited that they seem to be drawn with a brush by nature, but as they are not as good breeders as the sparrows of Japan (which without the exuberant beauty of the previous ones) are nevertheless excellent mothers, they will be the ones used as wet nurses to carry out the breeding of the Gould's Diamonds.

To start breeding Gouldian Diamonds, a few pairs are enough if you take into account the characteristics of their hereditary transmission, so you can easily achieve an immense variety <u>by using 4 or 5 pairs of</u>

<u>wet nurses for each pair of Lady.</u>

Before starting with the different varieties we must point out that Gould's Diamonds are one of the rare species where the "wild" is represented by three kinds of individuals, the red-headed, the black-headed and the orange-headed that live more or less mixed and cross without difficulty between them to give individuals of the three types, although the orange-headed because of their autosomal recessive character disappear fairly quickly from the population when in free mating with the red or black-headed that are transmitted with dominant character.

The black-headed variety was first discovered while the red-headed variety was later discovered by John Gould who is actually recorded in history as the discoverer of the Gould Diamond.
JUVENILE PLUMAGE
At the time of independence, Gould's Diamonds present a rather dull, dull, colorless and lackluster plumage.
The juvenile molt takes place in the first year, generally at two and a half months and lasts three months, providing the chicks with a fairly neutral plumage that will be preceded by the adult molt, one year later, which will provide the beautiful definitive plumage.

THE GOULD DIAMOND CLASSIC (Wild)

The "wild" or "base variety" is represented (we repeat) by three kinds of individuals, the red-headed, the black-headed and the orange-headed, which live more or less mixed and cross without difficulty with each other to give individuals of the three types, although the orange-headed, being of autosomal recessive character, disappear rather quickly from the population, being in free mating with the red-headed or black-headed, which are transmitted with a dominant character.
The so-called classic show bird has the following characteristics. Red or black head, mauve breast and red spotted beak.

HEAD COLOR IN GOULDIAN DIAMONDS

Various mutations can modify the colors in such a way that they can be obtained:

By carotenoid modification, Gould's diamond with red, orange or yellow head.

By mutation of the melanin, Gould's Diamond with gray or white and brown head, the latter called "Cinnamon" but very rare.

It was quickly discovered that the red color of the head dominated over the black,

Since whatever the color of the head, they all have red beaks, it is indisputable that these genes (head color and beak) are not linked.

But it has also been shown that **two conditions are necessary for the color red to express itself**:

- That a factor prevents the deposition of black melanin.
- That another factor allows carotenoid deposition.

These elements should be considered when you want to have or keep orange or yellow-headed Gould's Diamond.

A black-headed Gould's Diamond can be a carrier of orange <u>but not </u>of <u>orange head</u>. For to be effectively orange-headed it must have both a red factor and an orange factor (presence of colored mask for these factors) and this Gould's Diamond is necessarily red-headed. A Gouldian Diamond that has two orange factors is necessarily orange-spotted beak. Its head will be orange if it has a red factor and black if it does not, in the latter case it is called a Gouldian Diamond with a black head and orange beak.

PARROW (GUTTATAS).

(Emblem Guttata).

It belongs to the Estrildidae family.

In our environment they have been called simply Guttatas but this is a bad habit since there are other Guttatas such as *Poephila guttata* (known in Cuba as Cebrita) and this tends to confuse.

In other latitudes it is called Spotted Diamond or Fire-tailed Diamond. It is a very resistant and elegant bird, which makes it a real adornment in any aviary.

- Its breeding in guardianship dates back to the first half of the 19th century.
- Its size is around 12cm.
- The head is gray and the beak is red.
- It has a black band from the eyes to the beak.
- The eyes are black with an orange-red ring around them.
- The gray neck is separated from the white ventral part by a black band that extends to the tail and is mottled with white.
- The back may be light brownish brown. The tail is black.
- The female lays between 4 and 6 eggs, which she incubates for 15 days and 25 days later the chicks leave the nest.
- Sometimes it is necessary to use wet nurses to produce offspring.
- The rest of its characteristics are similar to those seen for all passerines.

BIBLIOGRAPHY.

1.   Álvarez González, Viviana. 1994. En compañía de las aves ornamentales, Editorial Científico Técnica. Havana.

2.Álvarez Olivar, Esperanza. 1989. Parakeets of Australia, brochure.

3.American Budgerigar Society. 1995. An handbook for the novice breeder.

4.   Blanco Castro, Enrique, El canario opalino, p. 19 Rev. Cuba Ornitológica, 2000,

5.   Brunelli, Ricardo. 1996. El gran libro ilustrado de los canarios, Editorial de Vecchi.

6.   Castillo Chang, Miguel, El doble factor intenso en los canarios p,6-7 Rev. Cuba Ornitológica Year 2 No 4, 2003

7.   Chirino Pedraza, Roberto. 1991. Reproduction and feeding of the parakeet, (brochure).

8.Del Pino Luengo, Miguel. 1976. El periquito, Aedos Publishing House.

9.Domefauna. 1993. Parakeets, team of specialists, editorial de Vecchi.

10. Elisabetta Gismondi. 2002. Cocorite, Editorial de Vecchi.

11. Fragose Rosario, Beatriz, Manejo y alimentación de las aves, pag. 8-9 Rev. Cuba Ornitológica Year 1 No. 2, 2002

12. Gismondi, Elisabetta. 1999. The great illustrated book of parrots. Editorial de Vecchi.

13. Gismondi Elisabetta. 1995. Complete guide to colored canaries. Editorial de Vecchi.

14. Halaburda, T. 2001. Agapornis, ver y conocer, Editorial hispano europea.

15. Hernández Muñoz, Abel. 2000. Birds and you. Editorial Científico-Técnica. Havana, 81 pp.

16. Hernández Ponce, Casimiro, El canario de canto, su entrenamiento, p, 28.

- 29 Rev. Cuba Ornitológica year 2 No 4, 2003

17. Jacd C., Harris. 1999. Carolinas, Spanish-European Publishing House.

18. Minguez Mireia. 1994. The canaries, Editorial de Vecchi.

19. Noel Betancourt. 2000. El criador collection (brochures).

20. Pérez Beato, Octavio. 1987. Genetics of color in the canary.

Editorial Científico Técnica. Havana.

21. Rodríguez Guzmán, Manuel, El diamante de Gould y sus mutaciones. Unpublished

22. Rosemary, Low. 1996. Loros y afines, Editorial Omega. Barcelona.

23. Rutgers, A. 1995. Colored parakeets.

24. Silva Batista, Shirley, Gould's Diamond, Challenges and experiences Pag. 18- 19 Rev. Cuba Ornitológica Year 1 No.1, 2001

25. Silva Castillo, Manuel. 1999. Hereditary expectations in lovebirds, Editorial Sanlope.

26. Soto, Lucas. 1955. El canario y demás aves canoras, Editorial Sintes.

27. Stanislav, Chvapil. 1992. Aves de jaula, Editorial Susaeta.

28. Streeter. Ray. 1994. My parakeet. Editorial hispanoeuropea.

29. Uribe, F. 1993. How to breed exotic birds. Editorial de Vecchi.

30. Vins, Theo. 1996. Parakeets, Editorial Omega.

31. Vriends, Mathew M. 1988. Guide to cage birds, Editorial Grijalbo.

yes I want morebooks!

Buy your books fast and straightforward online - at one of world's fastest growing online book stores! Environmentally sound due to Print-on-Demand technologies.

Buy your books online at
www.morebooks.shop

Kaufen Sie Ihre Bücher schnell und unkompliziert online – auf einer der am schnellsten wachsenden Buchhandelsplattformen weltweit! Dank Print-On-Demand umwelt- und ressourcenschonend produzi ert.

Bücher schneller online kaufen
www.morebooks.shop

info@omniscriptum.com
www.omniscriptum.com

Printed by Books on Demand GmbH, Norderstedt / Germany